Wood Truss Roof Collapse
Claims Two Firefighters
Memphis, Tennessee

Investigated by: J. Gordon Routley

This is Report 069 of the Major Fires Investigation Project conducted by TriData Corporation under contract EMW-90-C-3338 to the United States Fire Administration, Federal Emergency Management Agency.

Department of Homeland Security
United States Fire Administration
National Fire Data Center

U.S. Fire Administration Fire Investigations Program

The U.S. Fire Administration develops reports on selected major fires throughout the country. The fires usually involve multiple deaths or a large loss of property. But the primary criterion for deciding to do a report is whether it will result in significant "lessons learned." In some cases these lessons bring to light new knowledge about fire--the effect of building construction or contents, human behavior in fire, etc. In other cases, the lessons are not new but are serious enough to highlight once again, with yet another fire tragedy report. In some cases, special reports are developed to discuss events, drills, or new technologies which are of interest to the fire service.

The reports are sent to fire magazines and are distributed at National and Regional fire meetings. The International Association of Fire Chiefs assists the USFA in disseminating the findings throughout the fire service. On a continuing basis the reports are available on request from the USFA; announcements of their availability are published widely in fire journals and newsletters.

This body of work provides detailed information on the nature of the fire problem for policymakers who must decide on allocations of resources between fire and other pressing problems, and within the fire service to improve codes and code enforcement, training, public fire education, building technology, and other related areas.

The Fire Administration, which has no regulatory authority, sends an experienced fire investigator into a community after a major incident only after having conferred with the local fire authorities to insure that the assistance and presence of the USFA would be supportive and would in no way interfere with any review of the incident they are themselves conducting. The intent is not to arrive during the event or even immediately after, but rather after the dust settles, so that a complete and objective review of all the important aspects of the incident can be made. Local authorities review the USFA's report while it is in draft. The USFA investigator or team is available to local authorities should they wish to request technical assistance for their own investigation.

This report and its recommendations were developed by USFA and by TriData Corporation, Arlington, Virginia, its staff and consultants, who are under contract to assist the USFA in carrying out the Fire Reports Program.

The USFA greatly appreciates the cooperation and assistance received from Director Charles E. Smith, Fire Marshal Hubert D. Crossnine, Captain Jeff Pickett, Safety Officer Joe Caldwell, and other members of the Memphis Fire Department.

For additional copies of this report write to the U.S. Fire Administration, 16825 South Seton Avenue, Emmitsburg, Maryland 21727. The report is available on the Administration's Web site at http://www.usfa.dhs.gov/

U.S. Fire Administration
Mission Statement

As an entity of the Department of Homeland Security, the mission of the USFA is to reduce life and economic losses due to fire and related emergencies, through leadership, advocacy, coordination, and support. We serve the Nation independently, in coordination with other Federal agencies, and in partnership with fire protection and emergency service communities. With a commitment to excellence, we provide public education, training, technology, and data initiatives.

 FEMA

TABLE OF CONTENTS

Wood Truss Roof Collapse Claims Two Firefighters
Memphis, Tennessee
December 1992

Local Contacts: Director Charles E. Smith
Memphis Fire Department
65 South Front Street
Memphis, Tennessee 38103-2498
(901) 527-1400
FAX: (901) 528-9506

Lieutenant Joe E. Caldwell
OSHA Safety Officer

Hubert D. Crossnine, Fire Marshal
2668 Avery Avenue
(901) 320-5460

Captain Jeff Pickett
Supervisor, Fire Prevention Bureau

Fire Investigator Allen Roberts
Board of Inquiry Members:
Division Chief Dewey R. Harris
Division Chief Richard S. Mosby
Division Chief Larry L. McKissack
Battalion Chief J. Harvey Herring
Lieutenant Joe E. Caldwell
Driver Jeffery A. Kuntz

Ellers, Oakley, Chester, and Rike, Consulting Engineers

OVERVIEW

The lightweight wood truss roof of a church in Memphis, Tennessee, collapsed on December 26, 1992, just seven minutes after the first units arrived at the scene of a mid-afternoon arson fire. Two Memphis firefighters later died from burns that resulted from being trapped under the burning structure.

The significant factors in this incident include the short time that a lightweight wood truss roof structure can be expected to maintain its structural integrity when involved in a fire and the lack of warning indicators of pending collapse. Firefighters must identify buildings with lightweight wood truss roof systems and know the proper tactics and strategy to employ when a fire involves this type of construction. Additional topics to be considered include the use of incident management systems, particularly with regard to operational safety and crew accountability, and the protection afforded by protective clothing systems.

SUMMARY OF KEY ISSUES

Issue	Comments
Structure	Lightweight wood truss roof collapsed without warning. Firefighters trapped in burning rubble.
Firefighters Fatally Burned	Two firefighters died from burns resulting from entrapment in burning rubble.
Protective Clothing	Burns could have been reduced with fire resistant station uniforms, turnout pants and hoods.
Personal Alert Safety Systems (PASS)	Members were not equipped with PASS units that could have helped rescuers fine trapped firefighters more quickly.
Pre-Fire Planning	Pre-fire planning is needed to identify buildings with lightweight construction hazards.
Incident Management	Inadequacies noted in incident management and personnel accountability.
Communications	Incident Commander was unable to communicate with companies over tactical radio channel.

A suspect was arrested and has made a statement concerning his involvement in setting the fire. The case is being prosecuted by the United States Attorney under the Federal arson statute, which can result in the death penalty when the death of a public safety officer is caused by arson. The Memphis Fire Division's Arson Unit was assisted by the Bureau of Alcohol, Tobacco and Firearms in the investigation and processing of evidence. The motive is believed to have been to cover-up a burglary.

THE BUILDING

The Pilgrims Hope Baptist Church is located in a sparsely populated area in the northern part of the City of Memphis. It is a small, single story brick structure, located on a residential street in an area of widely separated single family homes. Access to the neighborhood is restricted by narrow streets bordered by drainage ditches.

The main part of the church was constructed in 1974 and a section was added-on some years later. The original building was approximately 70 feet long by 40 feet wide (2,800 square feet). The addition is approximately 30 feet long by 48 feet wide (1,460 square feet) and is located at the front of the building. (Diagrams appear on the following pages.) The newer section has a lower roof line and the attics are separated by the brick wall that served as the original front wall of the church. The original building houses the sanctuary and worship areas, while the addition is primarily a meeting hall and dining area. Due to the distances between structures in the area, there were no external exposures to the fire building.

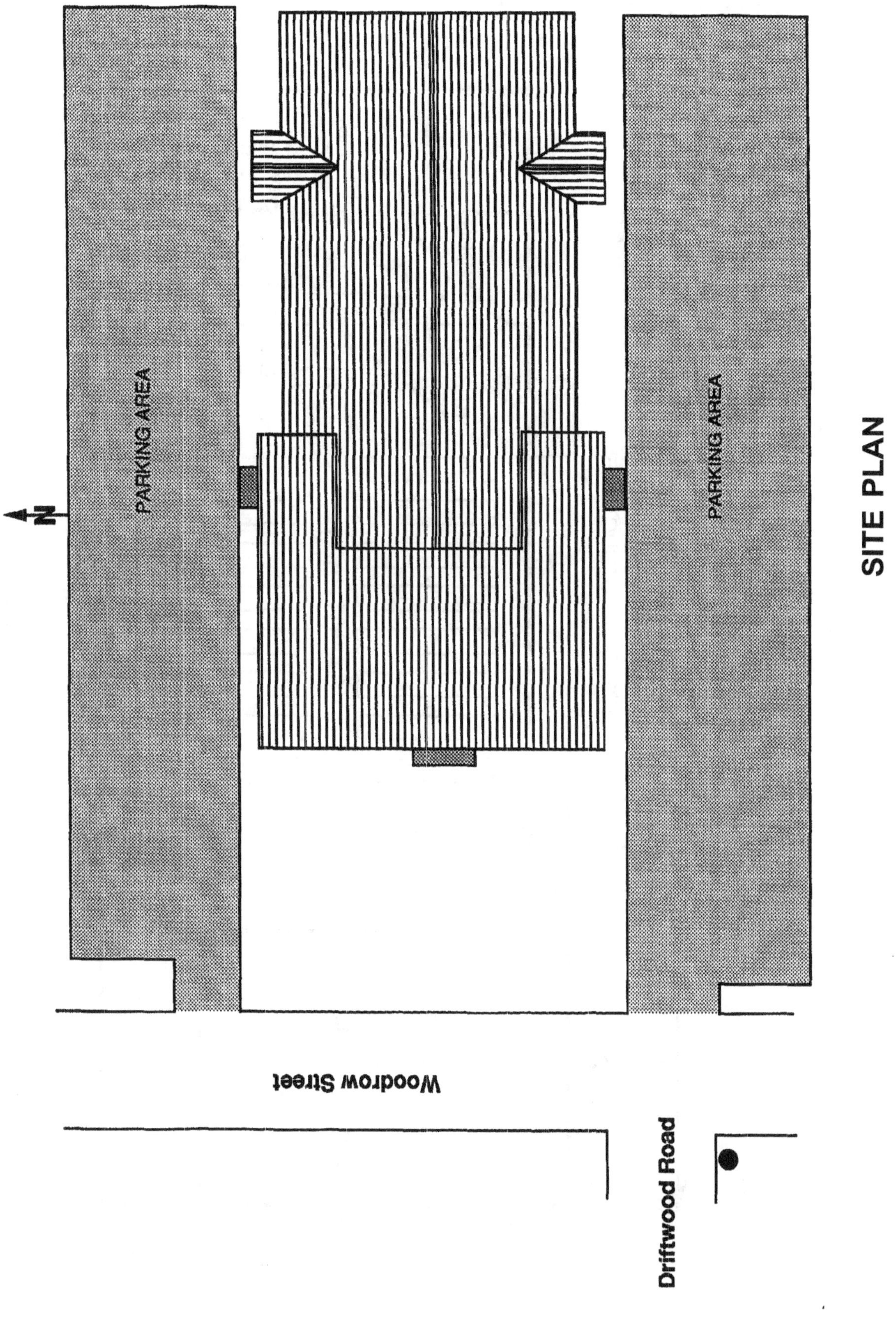

PARKING AREA

PARKING AREA

N

SITE PLAN

Woodrow Street

Driftwood Road

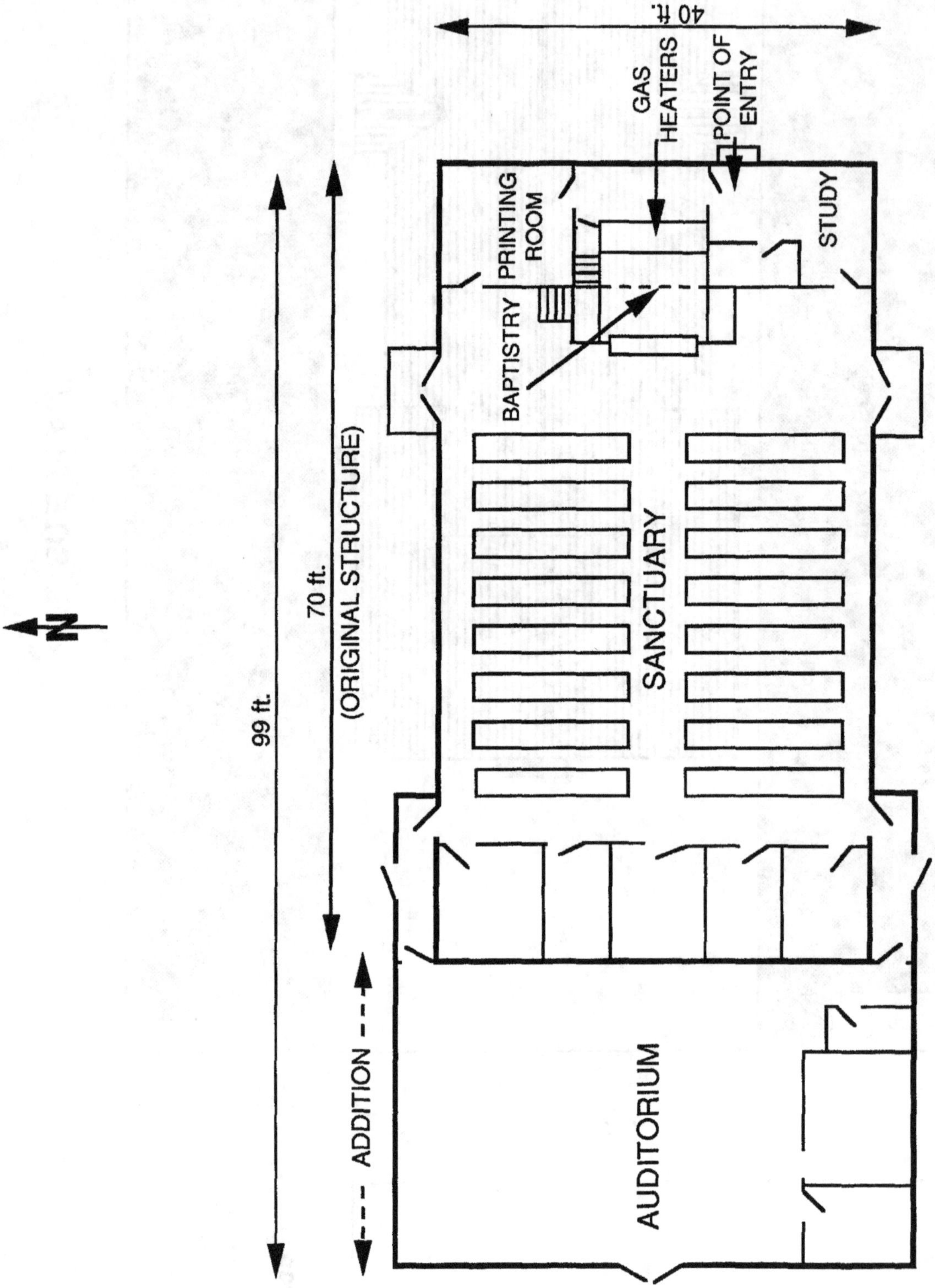

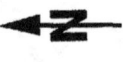

BUILDING PLAN

99 ft.

70 ft.
(ORIGINAL STRUCTURE)

40 ft.

ADDITION

GAS HEATERS

POINT OF ENTRY

PRINTING ROOM

BAPTISTRY

STUDY

SANCTUARY

AUDITORIUM

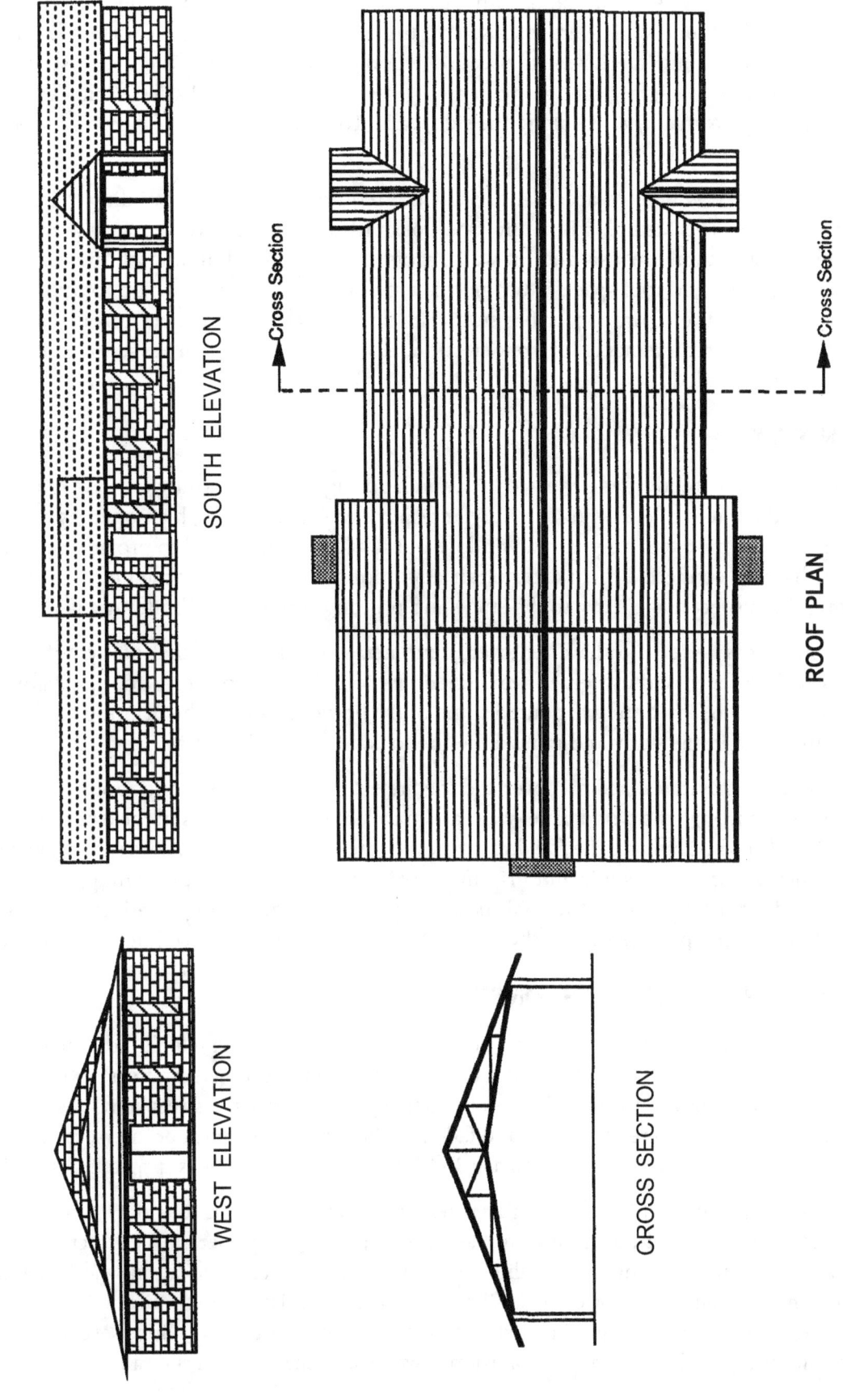

PILGRIMS HOPE BAPTIST CHURCH
Memphis, Tennesee

SOUTH ELEVATION

WEST ELEVATION

ROOF PLAN

CROSS SECTION

Cross Section

Cross Section

The construction was wood frame, with 2 by 4 wood stud walls and pre-engineered wood truss roof supports. The frame walls were enclosed on the exterior with a single course of non-load bearing brick veneer, and the interior finish was thin wood paneling nailed to the 2 by 4 studs. The gypsum board ceiling was slightly vaulted, with a height of approximately 12 feet 6 inches over the center aisle and approximately 8 feet along the side walls. The roof deck was 1/2-inch plywood covered by ordinary composition shingles.

The building was built in accordance with the requirements of the building code at the time of construction. The Memphis Building Code permits unprotected wood frame and has no requirements for structural fire resistance of walls, roofs, or floors for a building of this size and occupancy group. A concealed roof space (attic) in this type of structure may have up to 3,000 square feet of area without a fire separation or fire stop. The fire rating of the interior finish is also unregulated. There were no requirements for fire detection, alarm, or extinguishing systems in the building and none were installed.

IGNITION

The suspect apparently made entry to the church during the late morning or noon hour on December 26 with the intent of committing a burglary. Entry was made through an opening in the east (rear) wall. The opening was originally a window, but it had been modified to provide a passage for an air duct that was connected to an exterior air conditioning unit. After loosening a section of the duct work, the suspect made entry to the study and printing rooms. (See building plan on Page 5.)

The suspect is believed to have removed some items of value from the interior of the church and then to have been sniffing glue from containers found in the printing room. Before leaving the building he is believed to have poured some of the liquids found in the printing room over the floor and furniture in the study and ignited the fire to cover-up the burglary.

The room of origin reached flashover and the fire extended to the printing room. The flames quickly penetrated through the wood panel walls and traveled up in the spaces in the exterior walls, between the 2 by 4 studs. This gave the fire direct access to the entire undivided attic space. The wall dividing the sanctuary from the study and printing rooms extended only to the ceiling, so the fire was not restricted from extending to the void space above the sanctuary ceiling. Additional access to the attic would have been provided when the single thickness gypsum board ceiling failed over the fire area.

FIRE DEPARTMENT RESPONSE

The fire was reported by a passerby who called 9-1-1 on a mobile telephone. The caller reported a fire in a church, stating that the building appeared to be unoccupied and was on fire "all over." The caller could not provide a street address or intersection, because the street signs at the intersection in front of the church were weathered and faded; he had to drive to another intersection to read a street sign. The original call was answered at 1354 hours and the alarm was dispatched at 1357 hours.

The first alarm dispatch consisted of Engines 31, 26, and 19, Trucks 6 and 11, under the command of Battalion 11. All of the engine and ladder companies were operating with crews of one officer and three firefighters. Engine 27, which is normally first due at the location, was just going available from another call and reported its availability to respond. The Communications Center substituted Engine 27 for Engine 19 and directed Engine 19 to cover Station 27. Since several calls reporting a working fire were being received, Division 1 was also dispatched on the call.

FIRE SUPPRESSION OPERATIONS

Engine 31 arrived at 1402 hours, followed within less than a minute by Engines 27 and 26 and Truck 11. Truck 6 and Battalion 11 arrived two minutes later at 1404 hours. The acting lieutenant in charge of Engine 31, approaching from the north, assumed command and reporting a working fire with flames showing through the roof. Engine 31 dropped their "spaghetti load" (a 2 1/2-inch line wyed into two 1 3/4-inch attack lines) on the north side of the building and laid out to the corner of Driftwood and Woodrow where Engine 27 was already hooking up to the hydrant. (See figures on the following pages for conditions found on arrival and fireground operations diagram.)

The first attack line was extended by the officer and one firefighter from Engine 31 to the east end of the building where they found fire coming through a window opening. The second line was taken by the lieutenant from Engine 26 to the double door on the north side of the building near the east end. The firefighters from Engines 31 and 26 assisted with the two attack lines.

Truck 11 provided forcible entry to allow the second line to enter through the double doors on the north side to attack the interior fire. At this time the main body of fire was to the left of the entry point, in the smaller rooms behind a wall, and had extended only slightly into the sanctuary. The interior attack line was able to knock down most of the visible fire that had extended into the sanctuary in this area.

A larger body of fire could be seen extending into the sanctuary at the opposite (south) side, near the south entry doors. The area where the fire was through the roof was entirely to the east of the sanctuary wall. The fire had melted connections in two of three gas lines, which supplied unit heaters located behind the wall, resulting in a particularly intense fire at the point of burn-through.

Engine Company 27 hooked up to the hydrant and the crew extended a 2 1/2-inch attack line along the south side of the building. This line was initially operated through a window, where flames were visible, east of the double doors on the south side of the structure. When the doors were forced open this line was advanced inside to engage the fire immediately inside and to the right of the door. The line was advanced eight to ten feet into the building.

The crew of Truck 6 began to pull ceilings just inside and in line with the doorways on both the north and south sides. Some fire was encountered in the attic above them and both interior attack lines were operated through holes in the ceiling attempting to knock down the fire and prevent extension to the west. The crews operating these lines felt that they were making good progress at controlling the fire in the attic and did not believe that the flames had extended to the west of their positions.

When Battalion 11 arrived, two minutes behind Engine 31, the Battalion Commander attempted to contact the officer of Engine 31 and other company officers by radio. Unable to establish radio contact, he assumed that his radio was not functioning properly and set out on foot to make contact with the crews that were already working. The Battalion Commander met the officer of Engine 31 at the open door on the north side of the church. He assumed command of the incident and directed the lieutenant from Engine 31 to set up positive pressure ventilation (PPV) fans at the front of the building. The Battalion Commander also encouraged the hoseline crews to advance into the structure to attack the flames that were visible along the east wall of the sanctuary and in the rooms behind the wall.

The officer from Engine 31 went around to the west side where he encountered crew members from Trucks 6 and 11 already setting up their PPV fans at the front door. The fans were started, but a few

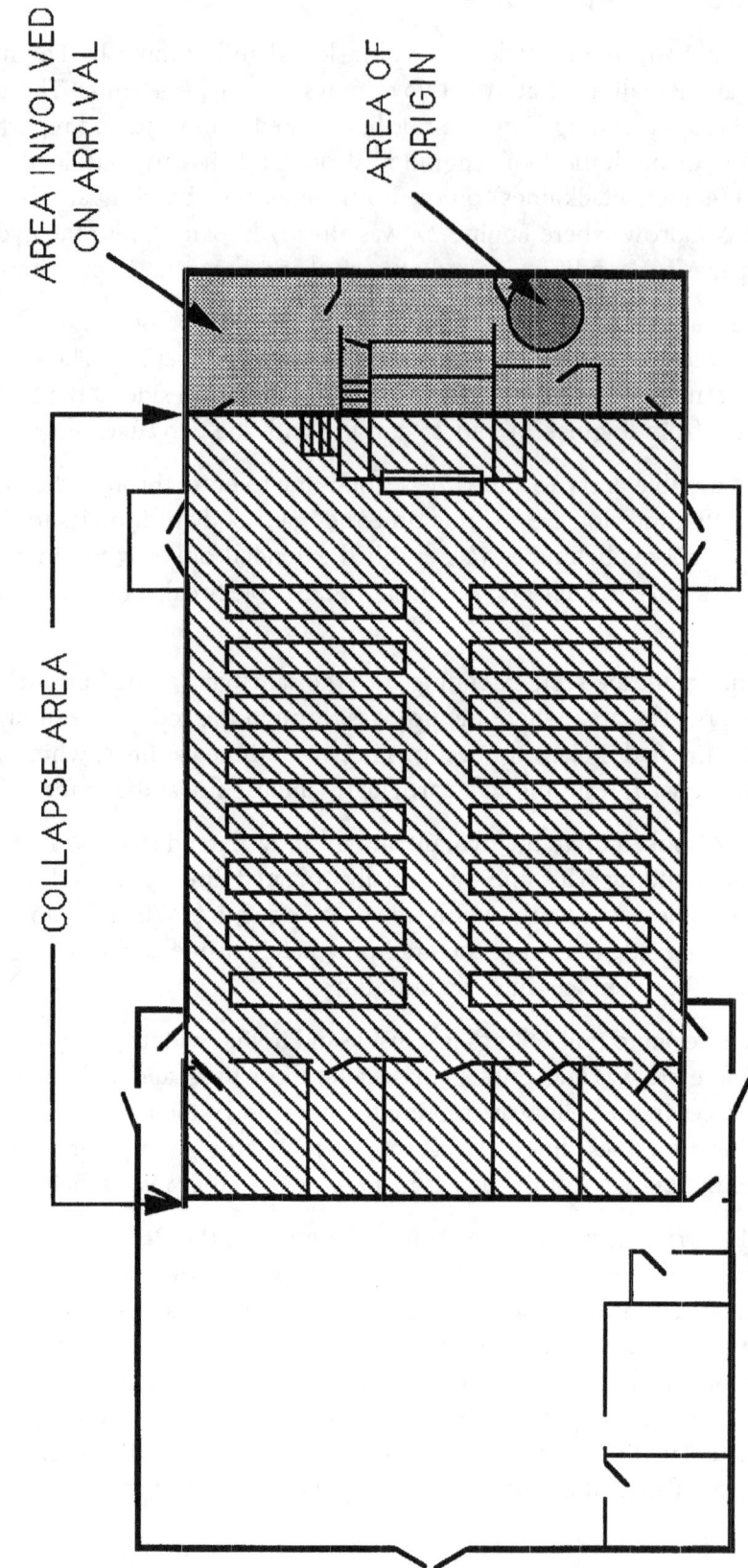

AREA OF FIRE INVOLVEMENT

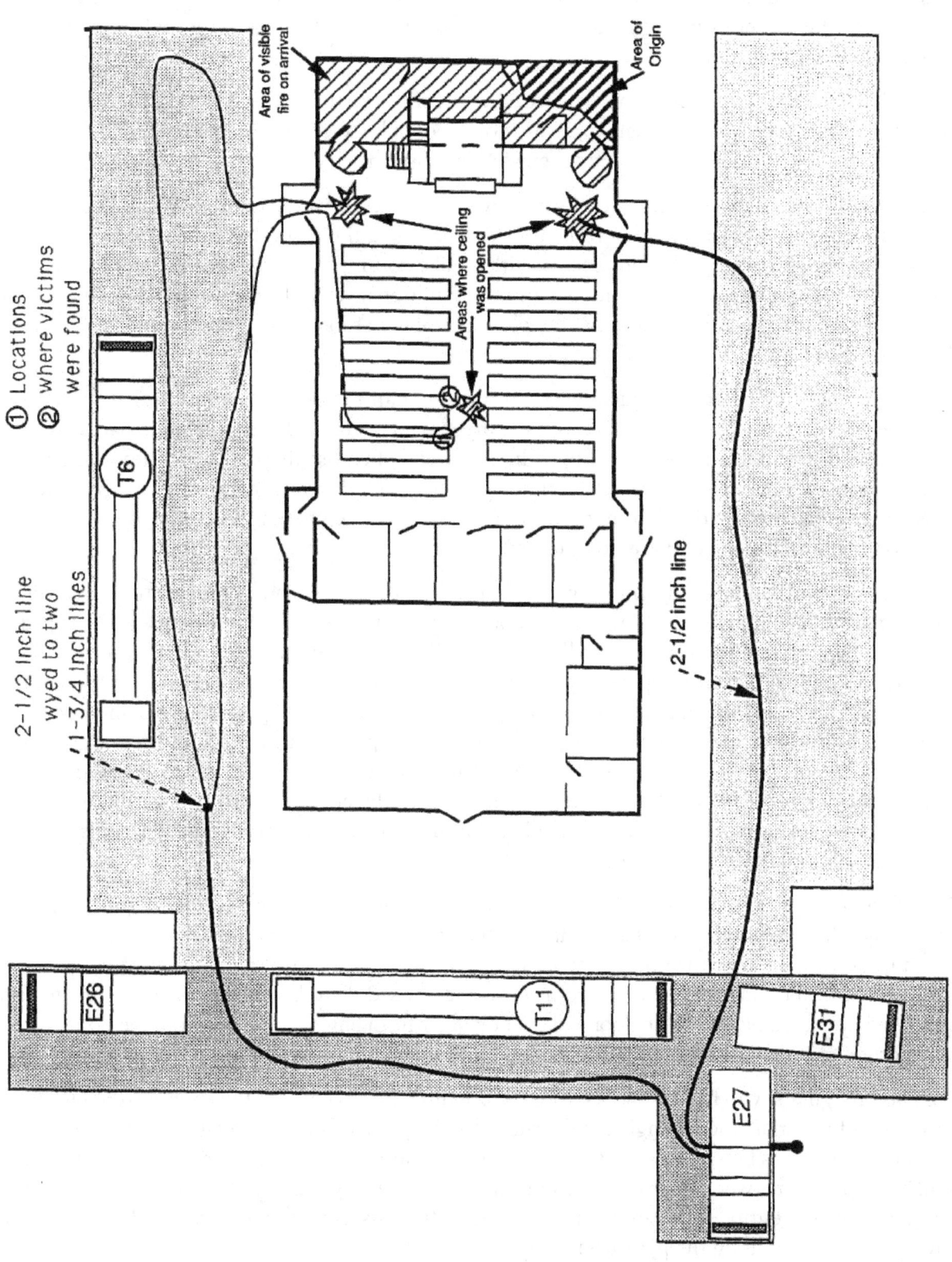

Locations ①
where victims ②
were found

Area of visible fire on arrival

Area of Origin

Areas where ceiling was opened

2-1/2 inch line wyed to two
1-3/4 inch lines

2-1/2 inch line

T6

E26

T11

E31

E27

seconds later the Battalion Commander came around to the front of the building and directed the personnel to turn the air flow away from the doors since the situation did not appear to require PPV at that time. The officer from Engine 31 then continued around to the south side and assisted Engine 27 with their 2 1/2-inch line.

When Truck 11 arrived at the scene, the acting lieutenant contacted the Battalion Commander by radio and asked if he wanted them to set up for a possible ladder pipe operation. The Battalion Commander directed Truck 11 to set up their PPV fan instead.

A designated crew member from Truck 11 went to the roof to assess the need for vertical ventilation. Noting that the fire was through the roof and free burning at the east end of the structure, and that the roof appeared to be lightly constructed with smoke coming from all openings, he determined that it would be unsafe to work on the roof and that additional vertical ventilation was not necessary. He returned to the ground level to assist the other members of his crew setting up the PPV fans and then pulling ceilings inside the church.

The Battalion Commander subsequently took a portable radio from the lieutenant of Truck 11 and attempted once again to direct operations on the fireground channel. He continued to have difficulty communicating with company officers by radio, apparently because they could not hear their radios over the ambient noise level on the fireground. (The fireground radio channels in Memphis are not repeated or monitored by the Communications Center; therefore, there is no tape or time log of the actual on-scene communications among units.)

The efforts to pull ceilings and control the fire in the attic at the east end of the sanctuary continued for several minutes. Personnel from all five operating companies were intermingled and had difficulty seeing each other in the smoke-filled interior. During this phase of the operation the 1 3/4-inch attack line that had been operated through the rear window from the east side of the church was brought inside by two firefighters from Engine 31 and one from Engine 26. This line was taken through the north doors to the center aisle of the church, approximately 25 to 30 feet west of the area where the other lines were being operated. A firefighter from Truck 6 assisted these members by pulling ceilings so that the line could be operated into the attic. The crews operating the other two lines were unaware that a third line had been brought inside and was being operated on visible fire in the attic, approximately 30 feet behind their position.

One of the firefighters from Engine 31, who had been operating the nozzle, became fatigued and went out through the north doors to rest after handling the nozzle to one of the other crew members. At the same time, one of the crew members on the 2 1/2-inch line, which had advanced several feet inside the south doors, felt burning debris falling from above and advised the other members on the line to pull back. As they backed out the south door and the firefighter from Engine 31 exited from the north side, a large portion of the truss roof caved in, dropping the peak of the roof down into the center aisle.

The two firefighters who had been operating the hoseline in the center of the church were buried in burning roof materials and entangled in the rubble of the partially burned trusses. They were pushed down into the spaces between pews at the north side of the aisle and enveloped by flames. All of the remaining personnel were close to the doors and were able to get themselves out from under the falling ceiling and escape to the exterior. The ends of the trusses remained supported by the walls, leaving a void space along the perimeter of the sanctuary.

RESCUE EFFORTS

The firefighter who had just left the two trapped members was the only one who was aware of their position and could direct the others to initiate rescue efforts. Several other members quickly became aware that at least one firefighter was trapped and began to direct their hoselines to the area where calls for help could be heard. The Battalion Commander was informed that a member was trapped and he directed Truck 11 to set up a ladder pipe to play down on the area, attempting to protect the trapped individual from the flames. Attempts were made to reach the trapped firefighter by making entry through windows and doors on the north side of the structure and by coming through from the front of the church.

One trapped firefighter was located, critically burned and entangled in the burning roof materials and the pews. He was extricated and carried out through a window opening to the north side of the church. Although critically burned, he was still conscious and able to talk when removed from the building. At that point it was realized that a second firefighter was missing and renewed efforts were made to locate and remove the second victim.

After the first victim was removed, it took several more minutes to locate and extricate the second victim. He was found about five feet east of location where the first trapped firefighter had been rescued, and rescuers had difficulty extricating him from the tangled debris. He was freed and was removed through a doorway to the south side. The second victim was also critically burned, but also still conscious and able to sit up and talk as he was treated and loaded into the ambulance.

The first victim was trapped for close to ten minutes and the second was not extricated for approximately 20 minutes, according to estimates of members on the scene. Both firefighters had suffered third degree burns to large areas of their bodies, including major respiratory system burns. They were transported to the Burn Unit at the Regional Medical Center, where they succumbed to their injuries on January 4 and January 11, 1993. Although they were conscious and able to talk when rescued from the building, their respiratory system burns rendered them unable to communicate with investigators within the first few hours.

FIRE CONTROL

Division 1 arrived at 1409 hours, approximately the time of the roof collapse, but was unaware of the fact that members were trapped for several minutes. He had to park several hundred feet from the scene due to traffic congestion and attempted to contact Woodrow Command by radio from his vehicle, but was unable to reach Battalion 11 on the incident channel. He then donned his protective clothing and, noting the amount of fire that was visible and the difficult access into the area, requested the response of two additional engines and one more truck to stage a block away. This request was made before he left his vehicle and was recorded at 1413 hours. He then set out on foot to locate the Battalion Commander. When he encountered the Battalion 11 he was told that the roof had collapsed and one firefighter was trapped inside – he could hear one of the trapped firefighters calling for help.

Division 1 assumed command of the incident at approximately 1417 hours. The fire was brought under control in approximately 30 minutes after the collapse, primarily by application of the ladder pipe stream from above the fire. The fire was confined to the original church area and the flames caused relatively minor damage at the floor level. Except for the area at the east end of the building, most of the fire involvement was in the attic before the collapse and in the roof materials after the collapse.

TIME SEQUENCE

The time of the collapse has not been precisely determined. Analysis of the various witness accounts suggests that the collapse probably occurred just as Division 1 was arriving on the scene, which was recorded by the Communications Center at 1409 hours.

Division 1, responding from the downtown area, initially observed a column of black smoke rising in the distance, suggesting a major free burning fire. As he approached the scene, the smoke had changed to a lighter color and diminished significantly, indicative of an effective attack being made on the fire. As he arrived, he observed a large body of fire coming from the center portion of the building. Since no witnesses reported seeing flames burning through the roof deck over the main part of the church before the collapse occurred, it is believed that Division 1 observed the fire that resulted from the collapse, although he did not become aware of the collapse and the fact that fire-fighters were trapped for at least five more minutes.

Division 1 called for an emergency medical services (EMS) unit to respond at 1417 hours, reporting that there was a firefighter trapped in the building and that crews were trying to reach him. This request was made shortly after he met up with Battalion 11 and was informed of the situation. Three minutes later, at 1420, he reported that the firefighter had been rescued and directed the EMS unit to come to the north side of the building.

The first EMS unit arrived and an additional EMS unit was requested at 1423 hours, when it was realized that a second firefighter was missing. A third EMS unit was requested at 1425 hours.

TIME TO COLLAPSE

The fire endurance characteristics of lightweight construction systems have been discussed and debated within the fire service for several years. There is considerable evidence to support the conclusion that these systems can be anticipated to fail quickly and catastrophically. The actual time of collapse and the presence or absence of warning signs prior to collapse are of great interest to the fire service.

This fire was well advanced before it was spotted and reported by the passerby. It cannot be determined how long the fire had been burning before it was discovered or how long the trusses in the collapse area were actually exposed to fire.[1] It is also not known if the flames had extended to the collapse area before firefighters arrived.

The fire originated in a room and extended to the attic via the void spaces between the 2 by 4 wall studs or through the ceiling. In either case the barrier was only a single layer of gypsum wall board which would have provided very limited resistance to the fire. The fire then involved the trusses and underside of the roof covering immediately above the area of origin. It is believed that the fire burned through the roof in this area before it was discovered. There is no way to determine if the fire spread laterally along the underside of the roof peak before it burned through the roof in the area of origin.

---

[1] It has been suggested that one of the initial actions that should be taken by firefighters arriving at a fire in a lightweight construction structure is to ask witnesses how long the fire has been burning. This fire is an example of a situation where it would be impossible to obtain a reasonable estimate of burn time or the time that the trusses had been exposed to fire before the arrival of the fire department.

The "Twenty Minute Rule" is used as a guideline by many fire departments as an indicator of the maximum time that crews should operate inside a burning structure, if the combustible elements of the structure are suspected to be involved. In this case the roof collapsed just seven minutes after the crews arrived on the scene and thirteen minutes after the first telephone call reporting the fire was received by the Memphis Fire Department. Inappropriate use of the "Twenty Minute Rule" in this situation could have led firefighters to believe that they had at least five more minutes to operate safely inside the building when the collapse occurred.

The crews working inside the building reported that they had no indication of structural failure until immediately before the collapse occurred. If there was any sag in the ceiling it was obscured by the smoke. The crews indicated that the visibility inside the church was so bad, due to smoke obscuration, that they had trouble determining the height of the ceiling.

The Incident Commander and most of the interior crews believed that they were effectively holding the fire from extending further to the west in the attic space, not realizing that the entire attic was involved. Only the two members who died, the firefighter who had just left them, and the firefighter who had pulled the ceiling for them appeared to be aware of the fire involvement in the attic over their heads. If the two members who died had any warning of the collapse, they did not have an opportunity to take action or warn others.

No one on the outside reported observing burn through or any signs of fire over the main area of the church prior to the collapse, including the firefighter from Truck 11 who had gone to the roof to determine the need for ventilation. While his statement indicated a feeling that it would not be safe to work on top of the roof, he did not indicate that collapse appeared to be imminent.

FAILURE ANALYSIS

The trusses were pre-engineered metal plate connected wood trusses, spanning a distance of 40 feet. The trusses over the main part of the church, where the collapse occurred, were scissors design (see diagram on the following page) to provide a slightly vaulted ceiling. The trusses over the adjacent areas were conventional peaked trusses with a horizontal bottom chord to create a flat ceiling. All of the trusses were spaced at 24 inch centers.

The trusses were shop fabricated from ordinary dimension lumber with 2 by 6 top chords; the bottom chords and web members were fabricated from 2 by 4 sections. All of the connectors were stamped metal gusset plates providing 3/8-inch penetration of the pointed ends into the wood. All observed connectors were fabricated with gusset plates on both sides of each connection point.

The attachment of the trusses to the 1/2-inch plywood roof decking maintained their alignment. There was no evidence of stringers or cross braces to maintain the spacing and redistribute loads among the trusses. Without cross bracing, the trusses were not constrained from roll-over, which can result in a "domino effect" collapse when a partial failure occurs. Cross bracing can allow a group of trusses to act as a system, with several trusses working together to support the load.

Cross bracing the trusses can reduce the probability of a sudden catastrophic failure and is more likely to result in visible sag before collapse. Without such load transfer mechanisms, the failure of any one truss can be sudden and catastrophic and can initiate the failure of a series of adjacent trusses in rapid succession, particularly if they are partially weakened and already close to their own point of failure.

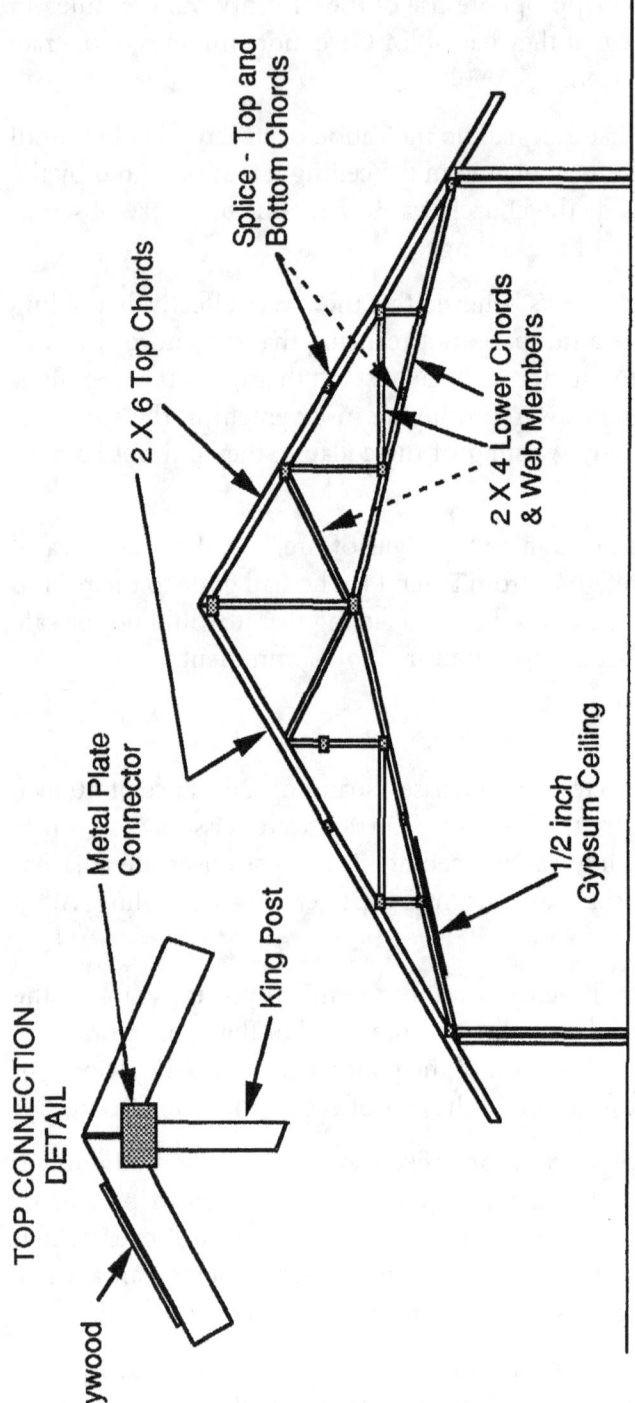

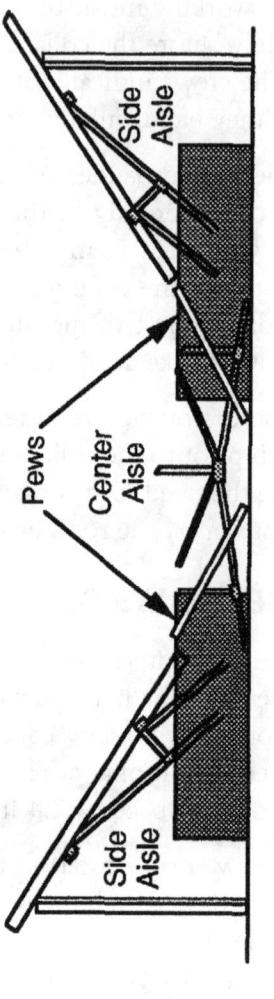

WOOD ROOF TRUSS DETAILS

The fire appears to have traveled along the underside of the roof, following the peak from east to west. The tops of all of the trusses were completely burned away for approximately 12 to 18 inches below the top connection point and could not be retrieved from the rubble. All of the lower connections could be located and most were still intact. The depth of char varied from almost full depth on the remnants of the higher members near the center of the span to almost no sign of burning on the lower and outer sections.

All of the top and bottom chords had splices near the mid-span points on each side of the trusses and most of these splices had failed, dropping the center section all the way to the floor, while the ends only rotated on their wall supports.

Many of the gusset plate connectors had failed, either by completely disengaging from the wood or by allowing one or more of the members to work free from a connection point. At the higher points the wood was charred beyond the depth of penetration of the gusset plate points, leaving empty pits in the charred wood where the gusset plates had been attached. As noted previously, virtually all of the top connection points were completely burned away. Some of the kingpost members were charred to a depth of more than 1/2-inch before the remaining unburned wood had broken; the diminished cross section of unburned wood fractured when it was unable to carry its load. (This would account for the "snap" sound reported by some of the firefighters just before the roof caved in.)

The depth of char on the roof materials decreased with distance toward the outer walls (downslope) from the mid-point. After the fire was extinguished, most of the plywood roof decking remained intact, except for approximately five feet on either side of the mid-point, still supported on top of the outer walls. The remaining unburned plywood was so thin, however, that there was obvious sag between trusses.

The roof collapse can be directly attributed to failure of the top connection points of the trusses, where the kingpost meets the top chord sections. It cannot be determined precisely which truss failed first and whether the gusset plates came loose from the top connection or the kingpost snapped and pried the gusset plates loose. The top connection is a critical point in each truss and its failure would result in the failure of that truss. It appears that several trusses failed in rapid succession – almost simultaneously following the first. All of the trusses would have been close to the point of failure and unable to support any additional load, so the load shift that occurred with the first failure probably caused several more trusses to collapse.

PROTECTIVE CLOTHING AND EQUIPMENT

Both of the firefighters suffered critical burns which ultimately resulted in their deaths. One survived in the burn unit for more than a week and the other for two weeks. The length of time they were buried in the burning rubble was the critical factor that resulted in the extent of their burns.

Both firefighters were wearing turnout coats, 3/4 length rubber boots, helmets, gloves, and self-contained breathing apparatus (SCBA). They had been issued fire resistive hoods but were not wearing them at the time. The Memphis Fire Department was in the process of transitioning to turnout pants, but the new clothing had not been issued to all members and its use had not yet been mandated. The firefighters also wore non-fire resistive station uniform pants and shirts that were made with a combination of polyester and cotton.

There are indications that the firefighters could have survived the experience if they had been fully protected and had been rescued more quickly. Both firefighters suffered major burns to the exposed areas of their bodies, particularly the head, neck, and thighs, and respiratory tract burns. They were both in critical condition for the entire period that they were hospitalized in the burn center and unable to communicate with investigators due to the respiratory tract burns.

Both firefighters still had their SCBA backpacks on when they were found. One had either removed his face piece or the face piece had been dislodged during the collapse, while the other still had his face piece in-place. Investigators were unable to determine whether or not the breathing apparatus continued to provide protection for a time after the firefighters were trapped, but they were trapped long enough that their air supplies should have been exhausted. A thorough examination of the breathing apparatus after the incident indicated that all components were functioning properly, except for components that were damaged by the direct fire exposure.

The polyester content of the station uniform clothing appeared to have contributed to the legs and lower torso burns, but the lack of turnout pants to protect this area was a more critical factor. The areas protected by their turnout coats were not seriously burned, in spite of being exposed to the fire for approximately ten minutes in one case and twenty minutes in the other case. The length of their exposure to the fire exceeded the times that protective clothing is designed to provide protection by a wide margin. The combination of external and respiratory tract burns was a fatal combination.

At the time of this fire Memphis firefighters were not equipped with PASS devices, which are designed to sound an audible warning if a firefighter is immobilized. If they had been equipped with PASS devices, it is likely that the two trapped firefighters could have been located and rescued much more quickly. Also, if one or both had been provided with a portable radio, they might have been able to call for assistance. Under an effective incident management system all interior crews should be in teams of two or more and at least one member of each entry team should have a portable radio on the incident channel.

INCIDENT MANAGEMENT

The lack of an effective incident management system and crew accountability contributed the time it took to rescue the firefighters. The two trapped members were from two different crews and neither company officer was aware of the location or function of the individuals before the collapse. The only people who knew that a line was being operated in the collapse area or that there was active fire in the attic in that part of the building were the two trapped individuals and the third firefighter who had just left them.

It took several minutes to make everyone at the scene aware of the fact that firefighters were missing, and it was not until one had been rescued that it was realized that the second was missing. When the word spread that there were firefighters missing several took independent action in attempts to locate and rescue them. The efforts were valiant, but not organized, according to the accounts of individual action.

Communications was also a problem, contributing to the lack of organization at the scene, since the Incident Commander was unable to communicate with company officers on the tactical radio channel. If the danger of an imminent collapse had been recognized, the warning would have been delayed because the Incident Commander could not communicate with the company officers and the company officers did not know where their crew members were or what they were doing. The capability for information to be immediately communicated to the Incident Commander and for the Incident Commander to be able to direct crews away from danger is an essential component of an incident management system.

LESSONS LEARNED

1. **Awareness and concern about the hazards of lightweight construction need to be increased throughout the fire service.**

   The failure to recognize that the church had a lightweight wood truss roof system and that the attic was involved above and beyond the area where most of the crews were operating is the single most critical factor in this incident. Several similar incidents have occurred with similar roof and floor systems and with other types of lightweight combustible construction systems. All firefighters must recognize that this type of structure can collapse suddenly and without warning after a relatively short period of fire involvement. The actual time to collapse cannot be predicted; therefore, it is not safe for members to operate above or below this type of structural system for any period of time, if there is fire involvement of the truss space.

   The only reliable way to be aware of lightweight wood truss construction is to pre-fire plan all structures where it is likely to be found and make that information immediately available to responding companies and command officers. The construction details often cannot be reliably determined from inside or outside the building, particularly when it is on fire. During pre-fire planning visits, if ceilings have been installed under a floor or roof support system, it may be necessary to look above the ceilings and in other structural areas to determine the construction details.

   The information on lightweight wood truss construction and on other types of lightweight construction that are susceptible to sudden and catastrophic collapse must be managed in a manner that will make it known to responding fire service personnel. This may be accomplished through pre-fire plan information carried on apparatus or accessible through a computer aided dispatch system. There must also be a system to ensure that the appropriate officers are made aware of the information on dispatch or while en route to the scene of the fire. Some jurisdictions have adopted a policy of requiring buildings with this type of construction to be marked with a distinctive symbol that is visible from the street.

2. **An objective, thorough internal investigation can reveal critical areas for corrective action.**

   The Memphis Fire Department conducted an exhaustive internal investigation into the circumstances of this incident. The results of the investigation, which included several recommendations for changes and additions to procedures and additional training, were made known to the entire department. A consulting structural engineer was engaged to perform a detailed analysis of the structure and the collapse, and the investigation looked at all aspects of the fire department operation on this incident.

   One of the conclusions of the Board of Inquiry (Board memberships noted on Page 1) was that the Incident Command System was not effective employed at this incident. Many of the tactical decisions were made by individual company officers, in the absence of a strategic plan and a strong command presence in the early stages of the incident. The companies were "free lancing," and there was no accountability for crews or individual firefighters.

   It was noted that three of the five company officers at the scene were acting lieutenants who may have been inadequately prepared for their responsibilities. The recommendations included additional training and more visible identification for acting officers.

Radio problems, possibly including failure of some of the company officers to monitor the radio or to hear radio traffic, also contributed to the incident management problems.

The need for additional training was also noted, particularly for members to recognize light-weight construction buildings, including truss roofs, and to be aware of the hazards they present. To be aware of construction hazards in existing buildings will require comprehensive pre-fire planning and a system to make critical information available to companies and command officers responding to incidents. The Board of Inquiry's complete list of recommendations follows.

RECOMMENDATIONS FROM THE BOARD OF INQUIRY

1. Aggressively pursue city ordinances that mandate advance notice to firefighters of any building within corporate limits with truss roof features.

2. Develop and implement officer training classes at the Fire Academy. Update training requirements for promotional candidates to include specialized training, prior to allowing "out of rank" supervision of fire companies. These classes should place emphasis on Incident Command, standard operating procedures, and firefighting tactics.

3. Establish criteria allowing only the top rated promotional candidates to ride in an "out of rank" capacity after completion of the officers training class.

4. Furnish and mandate wearing of "out of rank" identification for all personnel doing so. This could be accomplished by helmet colors or Nomex type vests to be worn in addition to fireground protective equipment.

5. Review and emphasize additional training requirements of building construction. Particular emphasis should be placed on truss structures, as well as any other type construction that might prove detrimental to the safety of firefighters.

6. More emphasis to be placed on the Fire Communications Bureau to transmit any additional information received to responding companies. This specifically pertains to information concerning type of structure, location of the fire, amount of fire or smoke reported, and number of calls received.

7. Emergency units (paramedic ambulance) need to be automatically dispatched whenever more than one hoseline is laid at any fire scene. Units may be released by the Incident Commander after determination of need.

8. An additional command officer needs to be automatically dispatched to all incidents of a known working fire in commercial or large structures.

9. Additional emphasis placed on the Safety Officer position being established at all emergency situations involving firefighters or EMS personnel. The Safety Officer to be designated and responsible for identifying unsafe conditions in the working environment.

10. Increased emphasis on fireground communication between company officers, Incident Commander, and Safety Officer must be stressed.

11. Renewed emphasis and strict enforcement of all personal protective equipment being utilized at emergency incidents.

12. System devised and implemented for total accountability of individual fire suppression or EMS personnel on the scene of any emergency incident.

13. Evaluation of possibility of recording fireground communications on Frequency Six of portable radios.

14. The need for better coordination with Memphis Police Department for traffic control is an absolute necessity. (The lack of traffic and spectator control greatly inhibited the abilities of the responding firefighters and EMS personnel to access the scene of this incident.)

15. An aggressive, but functional company level inspection and pre-fire plan program should be initiated. The information should be documented and maintained by each fire company.

16. Establish training classes and furnish appropriate supplies on all firefighting equipment to enable treatment of critical burns prior to emergency units arriving on the scene.

17. Immediate directive prohibiting 100 percent polyester clothing of any type being worn by fire or EMS personnel involved in any aspect of firefighting functions.

Memphis Fire Director Charles Smith endorsed all of the Inquiry Board Recommendations. Although Director Smith had been in office only 11 months at the time of this tragedy, he had previously initiated processes leading to the implementation of National Fire Protection Association Standard 1500.

New PBI bunker gear ensembles were being field tested for purchase at the time of this incident. PASS devices had been ordered but had not yet been delivered by vendors.

An Executive Order, by Director Smith, had been issued, prohibiting the wearing of 3/4 length rubber boots and requiring full bunker protection, effective January 1, 1993. The fatal incident occurred just one week prior to the implementation date of the Executive Order.